Narrative Of A Voyage To India

Of A Shipwreck On Board The Lady Castlereagh And A Description Of New South Wales

By

W. B. Cramp

Double 9 BOOKS

Narrative Of A Voyage To India
Of A Shipwreck On Board The Lady Castlereagh And A Description Of New South Wales
by W. B. Cramp

ISBN: 978-93-59952-63-5

Published by

DOUBLE 9 BOOKS

2/13-B, Ansari Road
Daryaganj, New Delhi – 110002
info@double9books.com
www.double9books.com
Tel. 011-40042856

ABOUT THE AUTHOR

W. B. Cramp changed into a 19th-century author whose work, "Narrative of a Voyage to India," gives a completely unique and valuable angle on the experiences of travelers at some stage in that generation. This e book is a compelling journey narrative that chronicles Cramp's journey to India, a land that had long involved Western explorers and adventurers. In "Narrative of a Voyage to India," Cramp offers readers with an in depth account of his travels, full of shiny descriptions of the landscapes, cultures, and those he encountered alongside the manner. His narrative displays the curiosity and fascination that Western tourists regularly held for the unique and various subcontinent of India. Cramp's writing is marked by its observational richness, shooting the nuances of daily life, the intricacies of Indian society, and the sheer grandeur of the landscapes. His storytelling not best serves as an informative file of his adventure however additionally invites readers to share in his feel of marvel and exploration. While W. B. Cramp won't be as broadly identified as some of his contemporaries, his "Narrative of a Voyage to India" offers a treasured ancient and cultural lens through which readers can glimpse the attraction and demanding situations of journeying to this charming a part of the arena all through the 19th century.

CONTENTS

SECTION I

THE AUTHOR'S DEPARTURE FROM ENGLAND—DESCRIPTION OF THE CEREMONY ON CROSSING THE EQUINOCTIAL LINE, AND HIS ARRIVAL AT MADRAS.

On the 8th or 9th of January, 1815, we proceeded, in the Princess Charlotte, Indiaman, to North-fleet Hope, and received on board our cargo. On February 28th, we sailed to Gravesend, in company with the Company's ships Ceres, Lady Melville, Rose, and Medcalfe, and arrived at the Downs on the 3d of March. Our dispatches not being expected for some time, we moored ship. Our time passed on very pleasantly till the 27th inst., when the weather became rather boisterous, and accompanied by a heavy swell. On the evening of the 28th, as the Hon. Company's ship Tarva, from Bengal, was rounding the Foreland, she struck on the Goodwin Sands, and was forced to cut away her masts to lighten her, and get her clear off. The Ceres drifted almost on board us; we slipped our cables, and with difficulty escaped the Goodwin Sands.

On the 1st of April the pursers joined their respective ships, and on the 3d we made sail with a fair breeze, and soon cleared the English channel. Nothing was now heard but confusion; the pilot having just left the ship, the hoarse voice of the captain resounded through a speaking trumpet, while the seamen were busy in making sail. We had a fine steady breeze till we made the Bay of Biscay, when we had a strong gale for three days.

After the hurry and bustle of the gale was over, we had a fine steady breeze; I then began to feel an inward pleasure, and to rejoice in the predilection I had imbibed from my earliest years.

We arrived on the equinoctial about eight o'clock in the evening of the 19th of April, when one of the oldest seamen is deputed Neptune; when he went into the head and hailed the ship in the usual form, Ship, hoa! ship, hoa! what ship is that? The chief officer replied, The Hon. Company's ship Princess Charlotte of Wales, and that he would be glad of his company on the morrow. Gladly would I have dispensed with it. On his quitting the vessel, as is supposed, a pitch cask was thrown overboard on fire, which had the appearance of a boat till lost to view.

The next morning, about nine a. m., Neptune hailed the ship again, when he was invited on board (from the head). On the fore-part of the gang-way and after-part of the long-boat, a boom was placed across, and a tarpauling was hung in form of a curtain, so that when they were in readiness they took it down, and the procession moved on towards the cuddy, twelve of the officers walking in the front, two by two with staves (broomsticks); next followed Neptune's car, (a grating with a chair covered with sheep skins) with Neptune, and his wife and child, (a recruit's child, as we had 250 on board, of his majesty's 46th regiment) Neptune bearing in his hand the granes with forks uppermost, and the representation of a dolphin on the middle prong, and Neptune's footman riding behind (barber) his carriage, dragged by the constables. The captain and officers came out to meet him, and presented him with a glass of gin, which was on this occasion termed wine. After the captain's health was drank, he desired them to proceed to business, and to make as much haste as possible; they then proceeded to the starboard gang-way, and Neptune placed himself upon his throne (on the boom, close to the long-boat and wash-deck tub) the slush tub being filled with balls, and lather made of slush, and the barber standing ready to begin his work with a razor made of a long piece of iron hoop well notched; the engine was brought on the quarter deck, and began to play, to force those below that had not crossed the line. I had not been long below before an officer from Neptune came to me, and demanded me, in his name to appear before him at the starboard gang-way, whose summons must not be disobeyed. On my arrival at the gang-way, the usual questions were asked me, whether I had been that way before? Without waiting for an answer they placed me on the wash-deck tub, and the barber rubbed me with the back of his razor and then let me go, upon my previously having given an order upon my bottle.

I had hardly got upon the poop, when one of the men was brought upon deck who was neither beloved by the men nor officers; they then placed him upon the tub, and asked him several questions, and while he was in the act of answering them, they thrust some black balls into his mouth, and then rubbed his face and neck over with lather, and scraped it in an unmerciful manner till the blood run in several places; they next pushed him into the tub of water and kept him under for the space of a minute, which tended to smart and inflame the wounds. It was at least a fortnight before he could wash himself perfectly clean; but now several more shared the same fate. The sun was setting fast before the amusements of the day were finished. The clouds presented the most beautiful appearance, and the rippling of the sea, together with the flying fish, scudding along the surface of the water, afforded the mariner a great field of thought. At so grand a display of the great and wonderful works of God, what mortal can be unmoved, or deny the existence of a being which nature herself proclaims!

The evening was very fine and beautifully star-light, and the moon shone with resplendent brightness. After the company had withdrawn to their evening refreshments, I amused myself with walking on the solitary poop. The sea appeared to be an immense plain, and presented a watery mirror to the skies. The infinite height above the firmament stretched its azure expanse, bespangled with unnumbered stars, and adorned with the moon '*walking in brightness;*' while the transparent surface both received and returned her silver image. Here, instead of being covered with sackcloth,[A] she shone with resplendent lustre; or rather with a lustre multiplied in proportion to the number of beholders.

> [A] I must be excused for the ideal extravagance of "clothing" this nocturnal luminary in "sackcloth," on adverting to that unlimited flight of poetic imagination, which speaks of "*Heaven peeping through the blanket of the deep.*" *Vide Shakspeare's Macbeth.*

Such I think is the effect of exemplary behaviour in persons of exalted rank; their course as it is nobly distinguished, so it will be happily influential; others will catch the diffusive rays, and be ambitious to resemble a pattern so commanding. Their amiable qualities will not terminate in themselves, but we shall see them reflected in their families.

My readers, I trust, will not wonder at my meditations on these sublunary objects, when they consider that they are the seaman's guide, and from them the greatest sources of nautical information are derived.

In the midst of these pleasing reveries, I was aroused by the ship being taken a-back, the watch being completely intoxicated, and it was only with difficulty that they could do their duty. Nothing material happened till our arrival at the Cape, when we experienced a severe gale for three days. The sea being heavy, she pitched her portals under water. We were running at the rate of ten knots per hour, under bare poles; and we soon after made the trade winds.

On the 23d of June we arrived in Madras roads; from the deck the view of the land has a magnificent appearance; the different offices have, to the beholder, the appearance of stone, and they are formed along the beach in a beautiful manner; they are built with piazzas and verandahs, and they extend about one mile along a sandy beach, while the natives parading along the shore, and the surf spraying upon the beach, gave the scene a very picturesque appearance. The surf beats here with so much violence that it is impossible for any ship's boats to land without being dashed to pieces.

On our making land we espied a small craft, called a kattamaran, making towards us; it was manned with two of the natives naked, except a handkerchief round their waist, and a straw round cap (turban) made with a partition in it to keep letters dry. This bark is made of three long hulls of trees, about ten or twelve feet in length, tied together with a rope so as to make in the centre a little hollow; they sit upon their knees in the centre, and have a long flat piece of wood, about five feet in length and five inches in width, which they hold in the centre, and keep continually in motion, first on one side and then on the other, and in that manner they force the kattamaran swiftly through the water.

It is very remarkable that these poor creatures risk themselves through the surf for a mere trifle, to carry letters for the different commanders to their respective vessels, at a time when the surf is at a dreadful height. When these poor fellows lay themselves flat on the kattamaran, and then trust themselves to the mercy of the surf, they are often driven back with great force, and they as often venture again, till they effect their purpose. They generally get their living by fishing, which is done by hook and line, and they offer them alongside the different ships for sale.

For two days the surf being so violent no boats could come off; but early on the third morning there were several came off with debashees (merchants) on board. They brought such things as might be wanted by the ship's company and officers. Their boats are made to carry passengers and cargo. There is not a vestige of a nail to be seen in them, their seams, instead of being nailed, are sewed together with coir rope; and they are generally manned with six or eight men.

SECTION II

THE AUTHOR'S DEPARTURE FROM MADRAS AND ARRIVAL AT BENGAL—DEPARTURE THEREFROM—HIS VESSEL RUNS ASHORE ON THE PULICAT SHOALS, AND GETS SAFE AFLOAT AGAIN, AFTER BEATING SIX HOURS AND FORTY MINUTES—HIS SAFE ARRIVAL AT MADRAS, AND DESCRIPTION OF THE DIVERS—ARRIVAL AT BOMBAY—THE SHIP BEING DOCKED, THE AUTHOR IS SENT TO BUTCHER'S ISLAND WITH THE SHIP'S COMPANY—A DESCRIPTION OF THE ISLAND OF ELEPHANTA—HIS JOINING THE SHIP AFTER HER LEAVING THE DOCK—HIS WORDS WITH HIS COMMANDER, AND BEING TURNED BEFORE THE MAST IN CONSEQUENCE—HIS DEPARTURE FROM BOMBAY, AND AFTER A SHORT PERIOD HE IS REPLACED IN HIS FORMER SITUATION—AND ARRIVES AT MADRAS.

We sailed from Madras, August 23d, and arrived at Bengal on the 30th. The scenery on the entrance up the river was indeed sublime, and inspired us with a sensation of gratitude to the Giver of all good. I went up to Calcutta with a craft of cargo; but having been sent down immediately, I could form no idea of the place.

On the 20th December we sailed from Bengal bound to Madras, in company with the Honourable Company's ship Marquis of Wellington. We kept a-head of her on the morning of the 25th, till she was almost mast down, and expected to bring-to about twelve o'clock in the Madras roads; but our expectations were greatly damped by the following circumstances:—At 8 A. M. the ship struck on the Pulicat rocks with such great violence, as to knock almost every man off his legs; the lead was immediately called, which, to the disgrace of some one, was not on deck; in the course of two minutes she struck again with as much violence as before; sail was immediately taken in, and after sounding, we found we drew about three and a half feet water. We then made signal of distress, by hoisting the ensign union downwards, and firing a gun. The Marquis of Wellington by this time hove in sight; all was confusion and consternation, the ship having beat several times with great violence. The Wellington hove to, and sent their cutter with four men and a second mate to our assistance, and then made sail and passed us, without rendering us any other assistance. The pinnace and long-boats, booms and spars, were immediately sent over the side, and the kedge-anchor was placed in the long-boat; but she leaked so very fast, that with all the united efforts of the seamen they could not keep her above water.

The weather was now very cloudy and black, and threatened a severe gale; so that our present situation became very disagreeable, as no assistance could be rendered us off shore, should necessity require it. But owing to the exertions of the officers and men, we effectually swung her head to the wind, which was blowing strong from the shore, and by 7 P. M. we anchored safe in the roads.

On the following morning we were busily employed in discharging our cargo and sending it on board its destined ships, (Honourable Company's ships Stratham and Rose.) After our clearance, the divers were expected from off shore, to examine the damage the ship's bottom had received; but, owing to the inclemency of the weather, it was impossible for them to get off from shore.

A seaman on board, by birth a West Indian, engaged to dive under the ship's bottom, and to acquaint us with the state of it, which was gladly accepted. In his youth he had been a fisherman on the coast of the island of Jamaica: the weather being rough, it was thought unsafe for him to venture; but on the following morning, it being quite calm, he prepared himself for his expedition: after he had jumped overboard, he walked, or rather trod water, round the ship; he informed us the copper was much battered above water, and in many places whole sheets of it were broken off; and after he had made us perfectly acquainted with the damages we had received above, he dived under her counter, and abreast of the after, main, and fore hatchways;—when he came on board, he informed us, that about twelve feet of our false-keel was knocked off, and about six feet of our copper abreast of the main-hatchway, besides a quantity of copper in different places, all of which we found to be true after we were docked.

We received considerable damage on board; the bolts were started from her side about three inches, and the main-beams sprung. Three days after he had dived, the captain came on board with two native divers, and several officers of the different vessels lying in the roads, to survey the ship. When they went under they brought up the same account as our man had first given. After about an hour's consultation, our ship was ordered to Bombay to be docked, it being the most convenient one for a ship of our burden. In a few days after we proceeded on our passage, and arrived in safety, keeping the pumps in continual motion during our passage.

The Island of Bombay is situated on the west coast of the ocean, and one of the three Presidencies belonging to the Honourable East India Company, and is in Lat. 18° 55′ N. and Lon. 72° 54′ E. of Greenwich. As soon as we had discharged all our cargo, and the ship was docked, the ship's company and officers were sent to Butcher's Island.

Butcher's Island is a small island situated about four miles and a half to the westward of Bombay, and is in circumference about one mile and a half, and has been a very formidable garrison. In the centre is a small fort and two barracks, the latter we took possession of for the ship's company. Soon after our landing on the island, a party of us went over to the Island of Elephanta.

The Island of *Elephanta* is about one mile and a half to the west of Butcher's Island, and is inhabited by 100 poor Indian families. It contains one of the most stupendous antiquities in the world: the figure of an elephant of the natural size, cut coarsely in black stone, appears in an open plain, near the landing place, from which an easy slope leads to an immense subterraneous cavern, hewn out of the solid rock, eighty or ninety feet long and forty broad, the roof of which is cut flat, and supported by regular rows of pillars, about ten feet high, with capitals resembling round cushions, and at the farther end of it are three gigantic figures, mutilated by the bigoted zeal of the Portuguese, when this island was in their possession. After spending the day very pleasantly we returned.

The Sergeant (an old invalid) who had charge of the fort, had a beautiful little garden; thither in the morning I frequently resorted, to enjoy one of the most charming pieces of morning scenery that I had ever witnessed.

> "Awake! the morning shines, and the fresh fields
> Call you; ye lose the prime to mark how spring
> The tender plants; how blows the citron grove;
> What drops the myrrh, and what the balmy reed;
> How nature paints her colours; how the bee
> Sits on the bloom, extracting liquid sweets."
>
> Milton's Paradise Lost.

How delightful this fragrance. It is distributed in the nicest proportion; neither so strong as to depress the organs, nor so faint as to elude them. We are soon cloyed at a sumptuous banquet, but this pleasure never loses its poignancy, never palls the appetite; here luxury itself is innocent; or rather, in this case, indulgence is not capable of excess. Our amusements for the forenoon were our nautical studies, and in the afternoon officers and men joined in cricket. In the evening, after my duty of the day was dispatched, and the sultry heats were abated, I enjoyed the recreation of a walk in one of the finest recesses of the Island, and in one of the pleasantest evenings which the season produced.

The trees uniting their branches over my head, formed a verdant canopy, and cast a most refreshing shade; under my feet lay a carpet of Nature's velvet; grass intermingled with moss, and embroidered with the evening dew; jessamines, united with woodbines, twined around the trees, displaying their artless beauties to the eye, and diffusing their delicious sweets through the air. On either side, the boughs rounding into a set of regular arches, opened a view into the distant seas, and presented a prospect of the convex heavens. The little birds all joyous and grateful for the favours of the light, were paying their acknowledgments in a tribute of harmony, and soothing themselves to rest with songs. All these beauties of Nature were for a while withdrawn. The stars served to alleviate the frown of night, rather than to recover the objects from their obscurity. A faint ray scarcely reflected, and only gave the straining eye a very imperfect glimpse.

The day following that the ship came out of dock we joined her. Our labours were now unremitted, to get her in readiness for sea. Amidst all our exertions it was impossible to give any satisfaction; our chief mate was very arbitrary, and vented his spleen upon the defenceless midshipmen, besides making the backs of the poor seamen sore with *starting*. Starting is a term used for rope's-ending a man, or otherwise laying a *Point* severely across their shoulders till they have not the strength to wield it any longer; a point is a flat platted rope, made for the purpose of taking in reefs, or otherwise to fasten the sail upon the yards.

At length my life became so truly miserable, that I was determined in my own mind not to endure it, if there was any possibility of avoiding it. For that purpose I wrote on board his Majesty's frigate, Revolutionnaire, for a situation, when Captain Wolcombe generously offered me one, provided I could get permission of Captain Craig to leave my present ship. I was at length forced to leave Bombay, through this and other circumstances.

On our arrival at Madras every preparation was made for receiving our cargo on board, which was speedily done, and in a short time was ready for sea.

SECTION III

THE AUTHOR'S DEPARTURE FROM MADRAS, DESCRIPTION OF A WATER-SPOUT—HIS ARRIVAL AT ST. HELENA AND DEPARTURE THEREFROM, ARRIVAL IN ENGLAND—JOINS HIS MAJESTY'S TRANSPORT SHIP, TOTTENHAM, BOUND FOR NEW SOUTH WALES—HER RUNNING ON SHORE IN THE RIVER AND PUTTING BACK TO DOCK—HE AFTERWARDS JOINS HIS MAJESTY'S TRANSPORT SHIP, LADY CASTLEREAGH. HIS DEPARTURE FROM DEPTFORD AND ARRIVAL AT PORTSMOUTH—HIS DEPARTURE THEREFROM AND ARRIVAL AT NEW SOUTH WALES.

As soon as our dispatches were in readiness, we proceeded on our passage for England; the morning was beautiful, and as the men were heaving up the anchor, my heart felt an inward sensation of joy and gratitude to our Creator, that he had been pleased to bring us so far safe on our voyage; we made sail with a steady breeze, and soon lost sight of land. After we had been at sea about two days, close on our weather-bow we observed a water-spout; when we first saw it, it was whole and entire, and was in shape like a speaking trumpet, the small end downwards, and reaching to the sea, and the large end terminating in a black thick cloud: the spout itself was very black, and the more so the higher up; it seemed to be exactly perpendicular to the horizon, and its sides perfectly smooth, without the least ruggedness where it fell. The spray of the sea rose to a considerable height, which had somewhat the appearance of smoke; from the first time we saw it, it continued whole about a minute, and till it was quite dissipated three minutes; it began to waste from below, and gradually up, while the upper part remained entire, without any visible alteration, till at last it ended in black clouds, upon which a heavy rain fell in the neighbourhood. There was but little wind, and the sky was otherwise serene.

On our rounding the Cape we experienced a very heavy gale, which continued for the space of ten days. We arrived at St. Helena in about ten days after clearing the Cape of Good Hope.

The approach to this Island is tremendous, it being an immense large rock in the midst of the sea, on which there is not the least appearance of verdure, houses, or indeed any sign of inhabitants, till you arrive at the anchorage, which is to leeward of the Island; and in turning round the corner of the rock is a fort, close to the water's edge, from whence they make all ship's heave to, till they have sent a boat on board from the Admiral; and in case no attention is paid to their signal, they fire a shot. After proceeding a little way, the town is discovered in the midst of a valley, and has a very picturesque appearance.

The produce of the Island is potatoes and yams. The yams are used in time of great scarcity of wheat, for bread; the inhabitants are under the necessity of boiling them 12 hours and baking them, before they can eat them; and in fact, many of the Islanders prefer them to bread. The coast produces an amazing quantity of fish, particularly mackarel, which are in great abundance, and run in shoals about six fathom under water. At this time Napoleon resided at Longwood.

After staying here 12 days, we proceeded on our passage to England, and arrived there in six weeks and two days.—The distressed state of England, and scarcity of employment determined me again to try my fortune abroad, and for that purpose I made several applications to the different owners, but for some time was very unsuccessful. At length I was engaged by Messrs. Robinson, to join his Majesty's Ship Tottenham, bound to New South Wales with 200 convicts. On June the 8th I joined her. After receiving all the ship's and government stores on board, we proceeded to Woolwich, and received on board 50 of our number, and in the afternoon of the same day we made sail, and on a sudden struck on a reef at low water; we were lying high and dry; every means was used to get her off, but without success, till we sent our convicts up to the hulks, and discharged our stores into the different crafts sent for that purpose, and by that means lightened her so, that at the flood she drifted; she was so materially damaged, it was deemed necessary she should return back to Deptford to Dock.

I had not waited long in London, before another vacancy occurred on board His Majesty's Transport Ship Lady Castlereagh, lying at Deptford, bound to the same Port. Shortly after I had joined her, we sailed to Woolwich, and received on board our guard, which was composed of a detachment of his Majesty's 46th regiment of foot, and after receiving a portion of our convicts, we proceeded on our passage to Portsmouth: we received another portion from Sheerness, and in two days arrived at Portsmouth. The remainder of our prisoners not being in readiness, we were forced to bring up and moor ship a cable each way.

Spithead is a spacious road for shipping, between Portsmouth and the Isle of Wight, and where they in general lie after they are in readiness for sea. I went on shore to see the town of Portsmouth. It is situated inland of Portsea; the streets are generally narrow, and rather dirty, owing to their not being properly paved.

The Dock-yards, as there are several, resemble distinct towns, and are under a government separate from the garrison. Here is a commodious arsenal for laying up cannon, and the fortress may be justly considered as the most regular one in Great Britain. The number of men employed in the different rope-yards generally is considered to be between eight or nine hundred, and the garrison is very large. The town of Portsmouth contains about 40,000 inhabitants, and the harbour is reckoned one of the finest in the world, as there is water sufficient for the largest ships, and is so very capacious that the whole of the British navy may ride in safety. The principal branch run up to Fareham, a second to Pouchester and a third to Portsea Bridge; besides these channels there are several rithes, or channels, where the small men of war lie at their moorings. Opposite the town is the spacious road of Spithead. On the 20th of December we received our convicts, and the following day we made sail and passed through the Needles, which are two sharp-pointed rocks at the N. W. end of the Isle of Wight, so called from their sharp extremities.

The prisoners, during their voyage, behaved themselves with great propriety, considering the variety of characters which we had on board. We arrived at New South Wales on the 26th of April, 1818, after a pleasant passage.

SECTION IV

DESCRIPTION OF NEW SOUTH WALES—DEPARTURE THEREFROM—ARRIVAL AT VAN DIEMAN'S LAND.

We now made for the eastern coast of New Holland, southward of Port Jackson; the coast has a most beautiful appearance, being constantly green during the year. From the south cape, about five leagues to the northward, is a most spacious bay with good anchorage, and sheltered from all winds. The natives are very ferocious; few vessels put in without partially suffering by their depredations, particularly seamen who, having ventured from their parties, have been by them cut off, robbed, and murdered. This place is called Two-fold Bay; ten leagues farther north is Bateman's Bay. Here is good anchorage and plenty of fresh water, but it lies open to the E. N. E. winds, and when they prevail they are accompanied by a heavy swell, so that it is impossible for vessels to lie secure. Seventeen leagues farther north is Jervis's Bay, and an excellent harbour and good shelter from all winds, with a fine sandy bottom. Round two small islands, at the mouth of the bay, there are two very large kinds of fish, which are caught in abundance with hook and line, called king fish and snappers.

The next harbour to the northward is Botany Bay, which is a capacious bay, with excellent anchorage for shipping; but the entrance is very dangerous to those commanders who are strangers to the coast. At the head of the bay is George's River, which extends about sixty miles up the country, and is navigable for small vessels of about 40 tons burden; on the banks of this river there are several settlements, which I shall hereafter describe. Nine miles farther north are the heads of Port Jackson; on approaching the heads from sea, the entrance is so narrow, and the rocks so perpendicular, that the opening is not perceivable at a distance.

On the south head is a look-out house, and a flag staff, on which a yellow flag is hoisted on the approach of any vessels from sea, which is answered by another signal staff on a battery at the north end of the town, called Davis's Point Battery, which is to be seen from all parts of the town, so that a vessel is known to be approaching before she enters the port. After entering the heads, the river runs due south for six miles, it then turns short round a point of land on the north shore, called Bradley's Head, which runs due west for twenty-four miles. After rounding Bradley's Head, the town of Sydney is perceivable, about three miles distant on the south shore. The anchorage is a small cove, as still as a mill-pond, land-locked around on all sides; the principal buildings in view are the stores and dwelling of Mr. Campbell, a Bengal merchant; they are built of white stone and have a noble appearance: the next is the government stores, a large stone building, at the end of which is the hospital, wharf, and stairs, the only public-landing place in the cove; here are two centinels continually parading the quay. From the landing place is a fine wide street, called George Street, with several fine stone and brick buildings, extending a mile and a half long, and joining the race ground. The public buildings in this line are the governor's secretary's office, an orphan school for female children, and the military barracks, with many fine private buildings, shops, &c. On the S. E. side of the cove is the government house, a low but very extensive building, surrounded with verandahs, and built in the eastern style, with an extensive park and garden surrounded with a high stone wall. About a quarter of a mile south of the government house is the general hospital, a large and extensive building, erected without any expense to government, the whole having been completed and paid for by three private gentlemen of the colony, for the grant of certain privileges. One mile further S. E. is Wallamolla, a fine brick and stone mansion, the property and dwelling house of John Palmer, Esq., formerly Commandant-general of the colony.

Between the general hospital and Wallamolla is the race ground, a fine level course three miles long, planned and laid out after the model of Doncaster race course, by order of his excellency Lochlin Macquarie. The races commence on the 12th of August, and last three days, during which time the convicts are exempt from all government duties. Convicts that are placed in the town of Sydney are in many respects happier than those farther inland; those who are employed in the service of government are under the inspection of the superintendent of the public works; they assemble at the ringing of a bell, in the government-yard, soon after day-light, and are mustered by their respective overseers and conducted to their work by them, having received their orders from the superintendent on the preceding evening. The overseers are themselves convicts of good character, and perfect masters of their different trades. They labour from day-light until nine o'clock, and they have then one hour allowed them to breakfast, then they return and work till three in the afternoon, and from that time they are at liberty to work for whom they think proper.

On leaving Sydney, the next settlement is Rose Hill, or, called by the natives, Paramatta, and it is situated due west up the river. Between Sydney and Paramatta there is but one settlement, about half way, which is called

Kissing Point, and close on its banks is a large farm, kept by Mr. Squires, who likewise carries on an extensive brewery. The principal edifice at Paramatta is the government stores, a large stone building; close to the landing-place, and leading into the town, is a street about a mile long. They are generally small cottages, and are mostly inhabited by the convicts; and to each is attached a small garden, which they are compelled to keep in good order.

There is also a large manufactory of flax, the produce of the country, of which they make coarse cloth of different descriptions. This town is under the direction of the bishop of New South Wales (Samuel Marsden) and is the place where the noted George Barrington resided many years as chief constable, and died in the year 1806, highly respected by the principal men of the colony. At eight miles distance, in a westerly direction, is the village of Galba, which is a very fertile soil, the farms being in high cultivation, the ground clear of timber, and numbers of sheep and oxen seen grazing in its fields. Two miles south of Galba is the village of Castle Hills, in appearance resembling Galba; and a number of farm houses scattered about as far as the eye can reach. About fourteen miles, in a S. E. direction, is the town of Liverpool, on the banks of George's River; here cultivation is making rapid progress; and on each side of the river are numerous farms, till the traveller arrives at its termination. From George's River a branch runs in a N. W. direction, is about twenty miles in length, and is called the Nepean River. Here the eye of the agriculturist would be highly delighted at the verdure that constantly appears in view; the farms are but thinly dispersed, as the Nepean is not navigable.

At the extremity of the Nepean is the most extensive tract of land that has yet been discovered. This tract is laid out in pastures, which are literally covered with wild cattle, the produce of six cows and a bull which escaped from the colony about forty years ago. They were discovered by a runaway convict, who returned to the settlement and reported his discovery, for which they pardoned him his crime of desertion. After leaving the cow pastures, due north is the town of Windsor, the most productive place in the colony for grain of every description, which is brought to be shipped on the River Hawksborough, in small crafts for that purpose. Windsor is sixty miles from Sydney, and the river is navigable all the way from the sea; its entrance is called Broken Bay, and is fourteen miles north of Port Jackson, and thirty miles north of Broken Bay.

The town of Newcastle is situated about seven miles up the river, called the Coal River, in consequence of coals being found there in great abundance, of very good quality. This town is a place where all are sent to that prove refractory, or commit any crimes or misdemeanors in the colony, and is much dreaded by the convicts as a place of punishment.

Newcastle is the last settlement to the northward of Sydney; the natives are black, and appear to be a most miserable race of people: they live entirely naked, both men, women, and children, and they possess not the least shame. They carry fish and game to the different towns and villages inhabited by the English, which they barter for bread, tobacco, or spirits; they are, in general, of a light make, straight limbed, with curly black hair, and their face, arms, legs, and backs are usually besmeared with white chalk and red ochre. The cartilage of their nose is perforated, and a piece of reed, from eight to ten inches long, thrust through it, which seamen whimsically term their spritsail-yard. They seem to have no kind of religion; they bury their dead under ground, and they live in distinct clans, by the terms Gull, Taury Gull, or Uroga Gull, &c. They are very expert with their implements of war, which are spears made of reed, pointed with crystal or fish bone; they have a short club made of iron wood, called a waday, and a scimeter made of the same wood. Those inhabiting the coast have canoes; but the largest I ever saw would not hold more than two men with safety.

Their marriage ceremony is truly romantic; all the youth of a clan assemble, and are each armed with wadays; they then surround the young woman, and one seizes her by the arm, he is immediately attacked by another, and so on till he finds no combatant on the field, and then the conquering hero takes her to his arms.

The different kinds of game which the colony produces, are several kinds of kangaroos, of the same species, but differing in size and colour. Beasts of prey have never been seen in the colony. The birds are, parrots, cockatoos, and a large one called *emus*, which have very long legs and scarcely any wings; they in general live upon fern, and weigh from seventy to eighty pounds; there are likewise a number of black swans. The woods abound with a number of dangerous reptiles, such as centipedes and scorpions.

Government not being disposed to receive all our convicts, we were taken up to proceed to Van Diemen's Land, with a crew of two hundred convicts, besides a detachment of one hundred and sixty rank and file of his Majesty's 46th regiment of foot. We sailed from hence, and arrived at Van Diemen's Land after a pleasant passage of six days.

Van Diemen's Land is situated south of the Cape of New Holland, and is a dependency under the control of the Governor-General. Here is a Deputy-Governor, who resides at the principal town, called Hobart's Town, situated about thirty miles up the Derwent; it is a town at present consisting of small cottages, or huts, built of wood, and with but few free inhabitants. The soil of the country is good; but there is a very inconsiderable trade. The Derwent runs ninety miles due west up the country. North of the Derwent, about twenty miles, is Frederick Henry's Bay, an immense deep bay, with good anchorage and shelter for shipping; and north-west of Henry's Bay is another fine river, called Port Dalrymple; it runs south-west ninety miles inland; at the head of it is a town, called Launceston; the inhabitants are principally convicts, and are employed in clearing the land for government. The native inhabitants of Van Diemen's Land are nearly the same as those of New Holland; and they at present hold no intercourse with the European inhabitants. After our prisoners were received on shore, they sent us another detachment of 150 rank and file of his Majesty's 46th regiment for Madras, and we began to prepare for sea.

SECTION V

DEPARTURE FROM VAN DIEMEN'S LAND AND ARRIVAL AT MADRAS—AN ACCOUNT OF A SEVERE GALE, AND THE GREAT DANGER OF SHIPWRECK, TOGETHER WITH HER WONDERFUL ESCAPE FROM IT, AND HER SAFE ARRIVAL IN CUDDALORE.

The morning was beautiful, and the noise of the crew weighing the anchor, created much life and bustle; and as we proceeded out of the harbour Nature seemed to smile, and bid us welcome to the watery element we had been so long traversing. A few days after, we entered the Endeavour Straits, which are about ten leagues long and five broad. We had several canoes off from the shore of New Guinea. It is a long narrow island of the South Pacific Ocean, and north of New Holland, from which it is separated by this strait, except on the north-east entrance, where it is counteracted by a group of islands, called the Prince of Wales's Islands. The land is generally low, and covered with an astonishing luxuriance of wood and herbage. The inhabitants resemble those of New Holland, omitting the quantity of grease and red-ochre with which the New Hollanders besmear their skins.

Their canoes are neatly carved, and are about twelve feet in length; they have outriggers to keep them firm on the water, and they are formed out of the hulls of trees; they carry about five or six men. They brought on board a quantity of shells, bows, arrows, and clubs, besides other trifling articles, and they would exchange with us for bits of old iron-hoops, or in fact any old thing, however trifling. The breeze freshening, we soon lost sight of the native merchants.

We arrived at Madras on the 12th of September, 1818, after a tedious passage. Owing to General Munro's intended departure for England, our cargo was immediately got ready, and as expeditiously received by us, and we were ready for sea on the 20th of October; but our dispatches not being in readiness, we were forced to remain at our anchorage, and on the morning of the 24th the clouds looked very black, and threatened a severe storm; but no preparations were made on board, and at 4 p. m. signal was made from the shore for all ships to leave the roads, which unfortunately was not noticed by many of the officers of the different vessels. At 5 p. m. the gale commenced; but through neglect the royal and top-gallant yards were not sent down, nor could the officer commanding be persuaded that any danger would arise from remaining at our anchorage; the ship's company now came aft and expostulated; but the officer in command called them all cowards, and said he would not start her anchor if it blew the masts out of her.

About 2 a. m. on the 25th, the gale commenced with the utmost fury, and she rode her scuttles under water, but as they were not secure, the sea came inboard and made very fast upon us. At 6 a. m. the water was three feet on the lee-side of our gun-deck, and from the continual working of the ship the chests broke from their fastenings.

After seeing a vessel go down at her anchors close on our starboard bow, the officer then gave orders for our cable to be slipped, which was immediately put into execution. John Gardener, a seaman, wishing to go aloft, and not taking proper hold, was blown from the rigging, and never seen again. We set the fore-sail, which immediately split; the mainsail, met with the same fate; the gaskets of the topsails gave way, and the sails split. At half past eight we found we had sprung a leak, owing to the ship's labouring so much; in the course of ten minutes we sounded, and found three feet water in the hold. The pumps were choaked; by 9 a. m. they were cleared, and by this time we had eight feet water in the well, and three on the gun-deck; the ship rolled very much, and the chests, guns, and water-casks, being all cast adrift, were dashing from larboard to starboard with the greatest fury. At 10 a. m. the ship labouring so much, and her being eight streaks of her main-deck under water, abreast of her main-hatchway, so that we had very little prospect of her living two minutes above water, it was thought necessary to send her mizen-mast by the board, in order to righten her; but while going, the mizen-mast heeled to windward and caught her royal-yards in the top-sail tye, and stayed her so, that we were compelled to cut away the main-mast, which carried the fore-top-mast and jib-boom; and, while in the act of going by the board, it knocked an invalid down and killed him on the spot. The ship rightened a little; but the sea was very boisterous, and we appeared to be in a valley in the midst of a number of tremendous high mountains, which to all appearance seemed ready to fall and crush us. The carpenter came forward, and informed us, that we had sprung another leak, and that we had ten feet water in the well; the men, as by one accord, dropped the pumps, and appeared to despair; we might all have well exclaimed with the poet,

> "Heaven have mercy here upon us!
> For only that can save us now."

"The atmosphere was hurled into the most tremendous confusion, the aerial torment burst itself over mountains, seas, and continents. All things felt the dreadful shock; all things trembled under her scourge, her sturdy sons were strained to the very nerves, and almost swept her headlong to the deep."

It would be in vain to attempt to give a description of our feelings at this critical moment, tortured as we were with anguish and despair. Every man seemed now as if all was given over for lost, when the carpenter came forward and informed us the leak was found out, and that with a little exertion it might be stopped; the men then rose with great vigour, flew to the pumps with renovated strength, and gave three cheers. The cabins were all washed down, and a party of men were busily employed throwing every thing overboard,—self was not considered,—the very last rag was committed to the furious elements without a sigh. At 11 a. m. the sea struck her starboard quarter-gallery and forced it from its birth, and as we were busily employed, a cry was heard, the starboard fore-mast port was carried away, and the sea forced itself with great rapidity along the deck; but the seamen flew to meet this new misfortune with the vigour of tigers, not considering the dangers they had to encounter, and thus effectually succeeded in stopping the leak.

While the seamen were busily employed, the troops were desired to pump, which they firmly refused, and said they would sooner sink, except a poor blind man, who could not keep from them; his reply was truly noble, and, I am sure, my readers will excuse my repeating it. "I am unworthy of the life I have if I do not exert myself in this hour of distress; if it has pleased God to deprive me of the blessing of sight, he has not of the feelings of a Christian." At half past eleven the gale greatly abated, and by this time the carpenter had stopped the leak, by using all the gunny bags and blankets that could be found; the damage was occasioned by the masts beating under her counter. By 12 a. m. it was a perfect calm; the men were now busily employed clearing the gun-deck, and securing every port-hole and scuttle in which they effectually succeeded by 1 p. m.

"For a moment the turbulent and outrageous sky seemed to be assuaged; but it intermitted its wrath only to increase its strength; soon the sounding squadrons of the air returned to their attack, and renewed their ravages with redoubled fury; and the stately dome rocked amidst the wheeling clouds. The impregnable clouds tottered on its basis, and threatened to overwhelm those whom it was intended to protect, the vessel was almost rent in pieces, and scarcely secure; where then was a place of safety? Sleep affrighted flew, diversion was turned into horror; all was uproar in the elements; all was consternation among us, and nothing was seen but one wide picture of rueful devastation.

"The ocean swelled with tremendous commotions; the ponderous waves were heaved from their capacious beds, and almost lay bare the unfathomed deep; flung into the most rapid agitation, they swept over us, and tossed themselves into the clouds. We were rent from our anchors, and with all our enormous load were whirled swift as an arrow along the vast abyss. Now we climb the rolling mountains, we plough the frightful ridge, and seem to skim the skies; anon we plunge into the opening gulf, we reel to and fro, and stagger in the jarring decks, or climb the cordage, whilst bursting seas foam over the decks. Despair is in every face, and death sits threatening in every surge." The whistling of the wind and roaring of the sea, together with the voice of despairing seamen, and the dreadful shrieks of the women, made us truly miserable; but we were forced to exert ourselves with assumed courage and vigour, which could only be imagined but by those placed in a similar situation,—our exertions were for life or death, knowing that if they once failed, that nothing was to be expected but to perish in a watery grave.

We kept the water under to about three feet during the time of this dreadful gale; about 4 p. m. it abated, and about 5 p. m. it blew a steady breeze from the south-west; and at 6 p. m. we went round her to examine the damage we had sustained; when, dreadful to relate, we found that a man and child had been washed out of their hammocks and perished; on proceeding along the waste we found two invalids had been jammed to death between two water-casks and the ship's sides, making a total of six lives lost during the storm.

The hatches were opened about 8 p. m.; but the provisions being so salt and sodden with the sea water, they could not be eaten, on account of the scarcity of fresh water. After the watch was set we laid ourselves down upon the upper-deck with no other covering than the starry heavens.

On the following day we commenced clearing the wreck, and rigging up jurymasts, which we happily effected before sun-set; and on the 28th we arrived at Sadras, which lay south by west of Madras, distant fifteen miles. We lay here till the 30th without any tidings of the captain.

The men from fatigue and pain, from sleeping on the wet decks, and continual pumping, came aft, and said the clouds threatened another storm, and that the monsoons were growing very strong, and in case the weather should alter for the worse, they had not strength left to work the ship in another gale, from want of nourishment; and that provided the officers did not think proper to remove to a place of safety, they were determined to take charge of her and proceed to Trincomalee, and deliver the vessel into the hands of the under-writers. All our remonstrances to them were in vain, until the chief mate pledged his word and honour, that if the captain did not join her the next morning, he would, ill as he was, take charge of her and proceed there himself.

On the following morning the captain joined her, with the hon. L. G. K. Murray, secretary to the board of trade at Madras, when they brought on board a quantity of provisions, which we stood very much in need of, and immediately made sail and arrived the same day at Pondicherry. The governor sent us on board a new anchor, as our own was sprung. Pondicherry is a town of Hindostan, under the French government, and situated on the coast of Coromandel, seventy-five miles S. S. W. of Madras.

On the following day we run into Cuddalore, a little above the first bar. Cuddalore is a town of Hindostan, one hundred miles S. S. W. of Madras. Thirty of the ship's company being sick, they, with me, were compelled to leave the ship, and forced to proceed on shore to the hospital. I was about this time seized with a violent fit of the cholera morbus. It is supposed to originate from the cold damp airs which are very prevalent at this time of the season. A gentleman's bungalow was humanely given up as a hospital, or friendly receptacle, for our incapacitated seamen, during our sojourn at Cuddalore.

The possibility of visiting the native town was precluded by the peculiar strictness of the regulations imposed upon us.

SECTION VI

THE AUTHOR'S DEPARTURE FROM CUDDALORE AND ARRIVAL AT PONDICHERRY—DEPARTURE THEREFROM, AND ARRIVAL AT MADRAS, WITH A DESCRIPTION OF THE SAME—ACCOUNT OF THE RELIGION, CUSTOMS, AND MANNERS OF THE NATIVES—DEPARTURE FROM MADRAS, ON HIS ROUTE TO NAGPORE,—ARRIVAL AT PONAMALEE, AND DESCRIPTION OF THE SAME—HIS DEPARTURE AND ARRIVAL AT CUDDAPAH.

After I had thoroughly recovered, through the interest of a young German widow, I obtained my acquittal from the ship, and then proceeded to New Town for my passport. New Town lies about two miles and a half E. N. E. of Cuddalore, and is the residence of the Europeans in that neighbourhood; the houses of the Europeans are generally built of brick and those of the natives of wood. The day after I had obtained my passport I proceeded on my route and arrived at Pondicherry the same evening.

Pondicherry is about four leagues in extent; the houses are built with brick, but the Indians use only wood, in the manner which we call lath and plaster. In a few days after I arrived in Madras, and took up my residence with a friend in Pursevaulkum.

A few days after my arrival I proceeded with my friend to town. Madras, or Fort St. George, is a fort and town of the peninsula, on the coast of Coromandel. It is the principal settlement of the English on the east side of the peninsula, and is a fortress of great extent, including within it a regular well-built city. It is close to the sea shore, from which it has a rich and beautiful appearance, the houses being covered with a stucco, called *chunam*, which, in itself, is as compact as the finest marble, bears as high a polish, and is equally as splendid as that elegant material. There is a second city, called Black Town, nearly four miles in circumference, separated from Madras by the breadth of a proper esplanade. Madras, in common with all the European settlements on this coast, has no port for shipping, the coast forming nearly a straight line, and being incommoded with a high and dangerous surf. The citadel is situated in the middle of the White, or English Town, and is one of the best fortresses in the British possessions. The town is also encompassed with a strong wall of the same stone as that with which the citadel is built, and is defended by bastions, batteries, half-moons, flankers, and mortars. Opposite the west gate of the citadel are barracks and a convenient hospital for the company's soldiers, and at the other end is a mint where the company coin gold and silver.

I was shortly after engaged as an overseer in the Madras Advertiser printing office, and as an assistant to the Madras Nautical Academy; but not agreeing with my employer I left it, and obtained permission to stop in the country as a free merchant.

Mr. M. R— —, with whom I resided, used all his interest to obtain for me some permanent situation under government, but it could not be effected. At length, being tired of an indolent life, I opened a school, which succeeded very well, when I was forced to relinquish it, owing to my ill state of health the confinement and severity of the weather brought on a languishing complaint, which would have terminated in my death had I persisted in continuing in my present employment.

My friend being obliged to quit Madras, left me and his brother in charge of his house. My friends, during his absence, greatly contributed to my amusement, and, in short, spared no expense. One morning, passing through Vessory Bazar, I was greatly shocked at seeing the nabob's elephant take up a little child in his trunk and dash its brains out against the ground; the only reason that could be observed was, that the child had thrown some pebble stones at it; and the only redress the poor disconsolate mother could obtain was a gift of fifty pagodas from the nabob, which is about equal to twenty pounds sterling.

During my friend's absence his mother and brother were carried off with the cholera morbus. The general estimate of deaths through the settlement is at least three hundred and fifty in one day; the natives have been known to sacrifice in one day and at one pagoda, fifty cocks and fifty kids, to appease their angry gods, and, in fact, some of the poor deluded creatures will go with a sword run through their cheeks in the fleshy part, and kept hanging in that position for some days, continually dance backwards and forwards through the different bazars; others have the palms of their hands pierced with a sword; others have their breasts burnt, and others again have an instrument run through their tongue in order to calm the wrath of their offended deities; nor can they, in their opinions, put themselves to sufficient torture.

Shortly after my friend returned, I went to reside with a friend at Royaporum, south of Black Town, and soon afterwards I was engaged as an examiner in the accountant-general's office. After I had been a short time in this employ, I received an order to prepare for my departure for Nagpore, in the service of his highness the Rajah. On my return from the Fort St. George, I was greatly surprised at seeing an old man standing with his bare feet upon two pieces of wood in the form of a pair of pattens, with pointed pegs uppermost; he stood in that position for several days, with the blood running in torrents, and several of those who passed by gave him what their circumstances could well afford. A few days after I was invited to witness an Hindoo ceremony. We took our station at the top of a rich Persian's house, opposite a spacious esplanade and contiguous to a large pagoda; in the centre of the esplanade was fixed a capstern, with a pole about sixty feet long, which was fixed so as to be occasionally raised or lowered. Shortly after our arrival, a native, decorated with flowers, proceeded slowly towards the pagoda with tom-toms, and all kinds of Asiatic music; after he had prostrated himself in the pagoda, the Brahmin, a kind of priest, struck his side with a leather thong till it swelled to a considerable size, and then forced a butcher's hook through his side; he then composedly walked to the machine, and suffered himself to be fastened to a rope and suspended in the air with no other support than the butcher's hook; he went at least three times round a circle of about one hundred feet, and he kept his arms continually in motion during the whole time, fencing and throwing flowers among the bye standers, which were immediately picked up by them and kept as a religious relic. This ceremony is performed yearly for the purpose of those who have lost their cast, and may regain it by voluntarily undergoing this treatment. Eleven of them went through this torturing ceremony.

I now began to put myself in readiness for my departure. On the morning of the 8th I dispatched my baggage and tents, together with a guard of eight peons (native police), which my friends had obtained for me, through their interest with the superintendent of the police. By the time I had taken leave of all my friends, and thanked them for their disinterested protection to a distressed seaman, I proceeded on my route (after receiving several more marks of their favours, Mr. C— — having presented me with an Arab horse, four baggage bullocks, and five hundred rupees, besides several letters of introduction) at eight o'clock in the evening. I travelled about five miles down the Ponamalee Road, and stopped at a village a little below the main guard, a small place with scarcely any fodder for the cattle. On the following morning, at a very early hour, we proceeded on our march, and arrived at Ponamalee about eight o'clock, where I found several of my friends waiting to take leave, as they expected that Ponamalee would have been the first stage.

After having taken farewell of each other they returned back to Madras, and I hired for the day a small bungalow (or garden house) opposite the fort, where I determined to stay. Ponamalee is about fourteen miles W. S. W. of Madras. This small and beautiful town is situated upon a rising ground, which commands an extensive view of the adjacent country. The number of Europeans residing here is but few, as it is entirely out of the road for traffic. There is a fort which is situated upon a rising ground, and gives the village a romantic appearance. It forms a complete square, and on each angle is a small place erected in form of the body of a wind-mill, which was used formerly for the purpose of solitary confinement when the troops were quartered here, but is now occupied as lumber rooms; the fort is garrisoned by pensioners. The grand entrance is on the south side, and a small wicket is usually on the west. The fort is surrounded by a large moat about thirty feet in depth, the water is very clear and good, and is drank by the natives. The inner part is far from being roomy, owing to the extreme width of the ramparts. There are two or three small buildings for the use of the commanding officers, but now the residence of a school-master and two sergeants; in the centre is a small building with a dome on the top, which was used formerly for a chapel, but is now converted into a school for the instruction of the poor soldiers' children, and the two barracks are occupied by pensioners.

On the following morning, about two o'clock, we prepared for our journey, and in a few days arrived at Naggery, a distance of about two hundred miles W. N. W. of Madras. The natives here are Hindoos, and the village is remarkably clean. The pagoda, or place of worship, is a fine large building, built in an oblong form, and beautifully gilt and carved all round with monkeys and apes. The Hindoos, in their manner of diet, are very abstemious, refraining from flesh; in fact, they will not eat any animal food; they are very regular in their morning ablutions, which they do by washing and marking themselves with chunam in the centre of their foreheads, according to the mark of their different casts. If any one neglects it he is immediately turned out of the cast, and his relations disown him, nor will they permit him once to enter their house. Such is their strictness, that the father has refused to see his son and the mother her daughter; and if they happen to perceive him at any distance they fly from him as they would from a serpent, thinking that his touch would pollute them.

The roads here are very bad, being principally jungle; their principal cultivation is paddy (a kind of oats). On my arrival at Nundihall I was determined to rest for a couple of days, as two of my servants were in a very ill state of health. Nundihall is a beautiful town, the houses are built of brick, and are generally from three to four stories high; the streets were very dirty, owing to the number of paddy fields that surround the city, as the growth of it requires that the earth should be completely covered with water. The natives are generally Hindoos and Moors. The town is surrounded by a high brick wall.

After leaving the town of Nundihall the roads were very bad, owing to the quantity of stones, and hills which were very steep and difficult to ascend. On the roads I had several disputes with the natives passing through Wuntimuttall, owing to my servants and the peons stealing the toddy from the trees. Toddy is a liquor which is extracted from the top veins of the cocoa-nut trees, which runs continually into a pot placed for that purpose. The liquor is very pleasant, and is reckoned very wholesome when drank early in the morning in a small quantity; if drunk in the heat of the day it causes acidity in the bowels, and often is the cause of the death of many Europeans. The natives drink it continually, and often get quite intoxicated with it.

We arrived at Cuddapah on the 21st instant; it is a large and commodious town, and is inhabited by Mussulmen. Cuddapah is situated N. W. of Madras, one hundred and fifty-one miles distant, and the general estimate of inhabitants is at about two hundred thousand. The principal houses are built of brick and the inferior ones of mud.

The Mahometans divide their religion into two general parts, faith and practice, of which the first is divided into six distinct branches—belief in God, in his angels, in his scriptures, in his prophets, in the resurrection and final judgment, and lastly, in God's absolute decrees. The points relating to practice, are prayer with washings, &c., alms, fasting, pilgrimages, and circumcision.

The Mahometans pray five times in twenty-four hours, viz.: in the morning before sun-rise, when noon is past and the sun begins to decline from the meridian, in the afternoon before sun-set, in the evening after sun-set and before day is closed, and again in the evening and before the watch of the night. They fast with great strictness during the whole month of Ramadan, from the time the new moon first appears, during which period they must abstain from eating, drinking, and all other indulgences, from day-break till night or sun-set.

The Europeans reside about two miles to the west of the native town, and have commodious houses, with fine spacious gardens; they are built of brick and much after the form of a gentleman's seat in England, but on a larger scale. I proceeded to the house of the collector, and on my road, my horse taking fright, I was thrown, and lost my purse containing all my money. My distress was now indescribable. Being left pennyless in the midst of a people totally destitute of Christian feeling, and without the probable means of obtaining the common necessaries of life, I arrived, in this miserable state of mind, bordering on despair, at the collector's, Mr. Hanbury, and after making him acquainted with my circumstances, he generously rendered me his assistance, paid my servants' wages that were in arrear, and kindly advanced what I thought sufficient to defray my expenses, having previously sent my peons back to Madras, and supplied me with fresh ones to proceed with me to Hydrabad.

On the following day the rain came down in torrents, accompanied with thunder and lightning, which kept me within my tent and caused me to exclaim with Dr. Henry, "O, ye lightnings, that brood and lie couchant in the sulphureous vapours, that glance with forked fury from the angry gloom, swifter and fiercer than the lion rushes from his den, or open with vast expansive sheets of flame, sublimely waved over the prostrate world, and fearfully lingering in the affrighted skies!" "Ye thunders, that awfully grumble in the distant clouds, seem to meditate indignation, and from the first essays of a far more frightful peal; or suddenly bursting over your heads, rend the vault above and shake the ground below with a hideous and horrid crack!" In the evening the weather began to clear up, which induced me to walk out, when taking two peons as a guard, I proceeded south of the town, on a beautiful plain: the pleasantness of the weather, and the stillness of the evening, tempted me to prolong my walk, and inspired my mind to contemplate on the wonderful works of Providence, who had so lately showered down his blessings upon me, in preserving me from want in the midst of a heathen world. The sun had almost finished his daily course, and sunk lower and lower till he seemed to hover on the verge of the sky!

The globe is now half immured beneath the dusky earth; or, as the ancient poet speaks, "is shooting into the ocean, and sinks into the western sea." The whole face of the ground was overspread with shades, and what the painters of nature call "dun obscurity." Only a few superior eminences, tipt with streaming silver, the tops of groves and lofty towers that catch the last smiles of day, were still irradiated by the departing beams. But, O how transient is the destination—how momentary the gift! like all the blessings which mortals enjoy below, it is gone almost as soon as granted. How languishingly it trembled on the leafy spire, and glimmered with dying faintness on the mountain's sable brow! till it expired and resigned the world to the gradual approaches of night.

SECTION VII

THE AUTHOR'S DEPARTURE FROM CUDDAPAH—DESCRIPTION OF THE DIFFERENT VILLAGES, AND ARRIVAL AT HYDRABAD—DESCRIPTION OF HYDRABAD, AND DEPARTURE THEREFROM—ARRIVAL AT NERMUL.

On the morning of the 27th, I proceeded on my route over the chain hills, with which the town of Cuddapah is surrounded; the roads are very good, but the steepness of the hills made it very fatiguing: in six hours I arrived at Batoor, a distance of twelve miles. Batoor is a large village, the houses are built of mud and bamboo, and form a motley group; the only protection they have from the number of robbers which infest that part, is a small fort, about two hundred square feet; the ramparts are about fourteen feet in thickness, and at each angle a small gun is mounted upon a pivot, about three feet from its walls; the fort in general is very much out of repair; the inhabitants are Hindoos, and are very indolent; the land is quite barren and free from cultivation. The cruelty with which Europeans in general act towards these poor captives is really disgraceful, and cannot but be censured by all who cherish the least trait of humanity with their breast.

When an European passes through any of the villages, and is in want of any coolies, or porters, to carry his baggage, he orders his guards to press every man he can meet with, and compel him to carry whatever his barbarous protector chooses he should labour under, and if there is not sufficient men, to press the women, without considering whether they have any family to provide for. It has been frequently known, that the mother has been forced to leave her infant babe from her breast upon the bare earth to provide for itself, to carry the baggage of a merciless enemy, whose only payment, after going fifteen or sixteen Indian miles, is, if she complains, a *bambooing*, (that is a caning,) and, perhaps, after she gets home, which cannot be till the next day, she finds her poor infant dead for want.

We passed through Parmunsa, and arrived at Moorkandah, which is a small village, and in a very ruinous condition, as it is at the foot of the Ghaut; the inhabitants are but few in number, and are principally Brahmins, consequently provisions are very scarce; on my requesting the cutwall, or headman of the village, to bring some fowls, he refused, and said there were none in the place, although I repeatedly heard the crowing of a cock. The impudent manner in which the man answered me, made me doubt the truth of what he said; in order to ascertain it, I took two peons and my gun and went round the village, and found a full grown cock; I caught it, and ordered it to be carried to my tent and killed; the natives by this time were in arms, and before any of us were aware of it, they had secured the peons and surrounded me, demanding the cock: when they were informed of its death, they all began to weep and raised a most lamentable cry, and said it was devoted to their god, and that the heaviest curses would follow me. I expected their denunciations would have paid for it; but in that I was greatly mistaken, for they demanded payment for it; and to avoid any injury to my peons, I offered them one rupee, considering that it would be equal to the price of eighteen cocks; but they disdainfully refused it, and said that they must offer gifts to their god to appease his anger, and to pay their sadura to intercede in their behalf. I remonstrated with them; but to no avail, as they would not take less than ten rupees. I tried all in my power to make my escape from them; but when they perceived my intentions, they drew their scimitars, and held them to my breast, and said, provided I did not accede to their offer, they would not spare the lives of my peons nor myself, as they could not get it replaced for forty times that sum, which was presented to them by their rajah. The price I considered to be extortionate, (but I paid it,) as fowls are sold in the different villages round that neighbourhood for one penny each, sheep for ten-pence, and every other article in proportion.

On the following morning, at a very early hour, I crossed the Ghaut; in the centre there is a very great declivity on each side the road, about two hundred feet in depth, and the Ghaut is very steep, and covered with flint-stone, which made it very difficult for the horse and cattle to pass: it is about twelve miles in length, and at the foot of it is the village of Badnapore. The inhabitants are very peaceable, and the village is close on the borders of Khristnah river. We made all possible haste to cross, which was effected by means of a large round basket, which is continually whirling round in the river. The river is about a quarter of a mile in width, but the heavy current carried us nearly two miles down; and owing to the exertions of the cattle, we encamped close on its banks. On the following day we passed Pungall-hill fort, which is situate on the summit of a very steep mount, and is built of mud, and large enough to contain ten thousand troops; it is only accessible on the north-east angle, which is easily blockaded in case of necessity. In five days we arrived at Hydrabad.

Hydrabad lies about 350 miles north-west of Madras; the houses are built of brick, and generally run four and five stories high. The inhabitants are principally Mahometans interspersed with Hindoos.

The Mahometans will not suffer a Christian to touch their cooking utensils or fuel by any means, and if such should be done, they consider them as polluted, and they will instantly break and destroy them; and while they are in the act of eating, if touched by any one of another sect, they will not swallow what is even in their mouth, but will throw it out, and go through a regular purification by washing and prayer.

After I had been at Hydrabad a few days, I joined a small party to view the interior: while we were taking breakfast, a cavalcade of elephants came up to the door with a number of peons. After we had mounted them we proceeded through the south gate into the city; the streets were particularly dirty, owing to there being no drains. The town is supplied with water by a well about two hundred feet in circumference.

On our entrance into the minister's house we were surprised at seeing a battalion of female sepoys (soldiers) presenting arms to us. We stood to see them go through their military manœuvres, which they did with dexterity; we then proceeded towards the house, which is built entirely of cedar-wood, but in a very ordinary manner, owing to the number of apartments: every room is carved in a beautiful and masterly style, from the ceiling to the floor. This ornament is very common among the lower classes, who have the devices of their gods carved on the doors of their houses. The apartments form a complete square, and in the centre is a stone tank. We next proceeded to a gallery of looking-glasses; the only one worthy of notice is about eighteen feet long and sixteen wide; there is likewise a whole length painting of Earl Moira, Governor-General of India. We afterwards proceeded to the palace of the Rajah: on our entrance into the inner court, we were agreeably surprised at seeing a quantity of tea-cups, saucers, &c. of various colours, placed against the wall in form of elephants, tigers, serpents, &c. in the most superb manner; in the centre is a large tank, containing a great quantity of salmon-trout. I had the honour of being introduced to the Rajah's sons, but his Highness was not present.

After having obtained a guard of twelve sepoys and two naigues, I proceeded on my route, and in a few days arrived at Nermul.

Nermul is a large and beautiful city, surrounded by a fort, and is about three miles in circumference, and is on a rising ground, 205 miles north-north-east of Hydrabad, and in the heart of the jungle, it is under the command of Major Woodhouse. The inhabitants are principally Moors.

I pitched my tent in the middle of a burying-ground, by the side of a running stream, and owing to the fatigue I had experienced, I now resolved to sojourn for two days. This place suited my present state of mind.

My attention was soon attracted by a magnificent tomb, and upon examining the inscription, it proved to be a rajah's. The gardens were ingeniously planned, and a thousand elegant decorations designed; but, alas! their intended possessor is gone down "to the place of sculls!"

While I am recollecting, many, I question not, are experiencing the same tragical vicissitude. The eyes of the Sublime Being, who sits upon the circle of the earth, and views all its inhabitants with one incomprehensive glance, even now behold as many tents in affliction as overwhelmed the Egyptians in that fatal night when the destroying angel sheathed his arrows in all the pride of their strength; some sinking to the floor from their easy chair, and deaf even amidst the piercing shrieks of their distracted relations; some giving up the ghost as they retired, or lay reclined under the shady harbour to taste the sweets of the flowery scene; some as they sail with a party of pleasure along the silver stream and through the laughing meads! nor is the grim intruder terrified though wine and music flow around.

"Those who received vast revenues, and called whole lordships their own, are reduced to half a dozen feet of earth, or confined in a few sheets of lead! Rooms of state and sumptuous furniture are resigned for no other ornament than the *shroud*, for no other apartment than the darksome *niche*! Where is the star that blazed upon the breast, or the glittered sceptre? The only remains of departed dignity are the weather-beaten hatchment. I see no splendid retinue surrounding this solitary dwelling. The princely equipage hovers no longer about their lifeless master, he has no other attendant than a dusty *statue*; which, while the regardless world is as gay as ever, the sculptor's hand has taught to weep."

SECTION VIII

THE AUTHOR'S DEPARTURE FROM NERMUL AND ARRIVAL AT NAGPORE—HIS DEPARTURE, AND ARRIVAL AT JAULNAH—THE AUTHOR'S DEPARTURE FROM JAULNAH AND ARRIVAL AT POONAH, AND DESCRIPTION OF THE VILLAGES WITH THEIR RELIGION—HIS ARRIVAL AT BOMBAY, AND HIS DISTRESS—SKETCH OF BOMBAY AND ACCOUNT OF THE PERSIAN RELIGION—HE JOINS THE HONOURABLE COMPANY'S SHIP MARQUIS OF HUNTLY, AS CAPTAIN'S CLERK—HIS DEPARTURE FROM BOMBAY AND ARRIVAL AT BENGAL.

After remaining two days, I proceeded on my route; and on the following day arrived at Wadoor, a distance of fourteen miles, across a long succession of hills, the roads over which are very rugged and covered with stones; Wadoor lies in a valley, at the foot of a large mountain, and is hardly perceivable from the top.

On the 20th December, we travelled along a beautiful and finely cultivated country, the produce of which is cholum and paddy, which grows in great quantities; the inhabitants are very civil, and principally Moor men. On the 25th December, 1821, I arrived at Nagpore, and on the same evening was seized with the Nagpore fever, which is always accompanied by fits of the ague. The fever is supposed to originate from the excessive heats of the day, and the extreme cold of the night.

I endeavoured as much as possible that my ill state of health should not keep me from my employment, but attended to it very assiduously; which I persevered in till the 27th of March, when the doctor informed me, that I had better leave the Presidency or I should endanger my life, as the hot winds generally set in in the middle of April, which frequently prove very dangerous to European invalids.

On the 2nd of April, after having previously obtained my passport and a guard of twelve Seapoys, I proceeded on my route, and towards evening arrived at Tukea, where, owing to my ill state of health, I was compelled to stop two days.

On the 12th I arrived at Ouronty, which is S. W. by W. of Nagpore, about 100 miles. The town is very large, and is surrounded by a brick wall; the houses are built of brick, and are generally three stories high. The inhabitants are Mussulmen. In the afternoon I went to the palace of the Rajah, (Rajah ram.) His palace outside is very dirty, owing to his guard making fires against the walls for cooking. On my desiring to see the Rajah, I was conducted through a long dreary passage, with the walls, to all appearance, covered with grease and filth, at the end of which is a large court-yard, which has a very different appearance, the Rajah's apartments being all round; at the end were six Peons waiting to conduct me to his highness, with silver staves, about eleven feet long, with a device of Mahomet on the top; on my introduction to the Rajah's apartment, he was sitting cross-legged with his hooker; at my entrance he arose and made three salams in token of respect to the British nation. After questioning me where I was going to, and my reasons for so doing, he presented me with two camel-hair shawls, by placing them across my shoulder; then taking his leave.

On the following day, I proceeded on my route, and on the 20th arrived at Luckenwarry; where there is good encampment and water, and the natives are principally Hindoos. Early on the following morning we began to cross the Luckenwarry Ghaut; the roads were steep and not above ten feet wide, and on each side a vacuity of about 250 feet deep. The light in the lantern being extinguished, and the moon being obscured, my horse, had it not been for the horse-keeper, would have precipitated me to the bottom; I instantly dismounted, and the horse-keeper led him till he was clear of the Ghaut. On the centre is a large gate, which stands about forty feet high, and which, during the war, had withstood a three months' siege.

Passing through the jungle between the villages of Currone and Chickly, we were greatly surprised at seeing a large party on camels; we hailed them and enquired who they were, but we could not by any means obtain an answer; when finding they persisted in their obstinacy, the Naigues suspected them of belonging to the party of Sheik Dullah, a noted robber who had already committed many depredations in that neighbourhood, and on our desiring them to move to the left of us if they were friends, they made a sudden halt; the sepoys then drew up in a line, and the followers began to guard their baggage, but when they saw our number, they went off to the left of us, grumbling.

On the 24th, we arrived at Jaulnah. It bears W. by S., of Nagpore, distant 180 miles. On the following day, after I had taken sufficient rest, I presented my passport to the Adjutant-General, and delivered up the guard, having previously obtained another. Jaulnah is a large town, surrounded by a brick wall, about twenty feet in height; the houses are generally of brick, and from three to four stories; the inhabitants are principally Hindoos, interspersed with Persians and Mussulmen. The cantonment is the head quarters of the British army on this side the Deccan.—Jaulnah has a civil and military government.

After staying two days, I proceeded on my route, and on the 19th of May I arrived at Poonah. It bears S. S. E. of Bombay, and is in the territories of the Peishwa: it is about forty miles distant from Bombay. I took up my residence with a friend, commander of the Sebundaries; during my route, I passed through Armigabad, Amednagur, and Seroor; which is the residence of Europeans, and has detachments of different regiments quartered at each town: their houses are in general of brick and stone, their religion is Hindoo.

The Hindoos are divided into four tribes, first the Brahmin; second, the Khatry; third, the Bhyse; fourth, the Sooders; all these have their distinct sects, and cannot intermingle with each other; but for some offences they are expelled their sects, which is the highest punishment they can suffer. In this manner a kind of fifth sect, called Pariah, is formed of the dregs of the people, who are employed only in the meanest capacity. There is a kind of division which pervades the four sects indiscriminately; which is taken from the worship of their gods Vishnou and Sheevah; the worshippers of the former being named Vishnou bukht, and of the latter, Sheevah bukht.

Of these four sects the Brahmins have the superiority, and all the laws show such a partiality towards them, as cannot but induce us to suppose that they have had the principal hand in framing them. They are not allowed the privilege of sovereignty; but are solely kept for the instruction of the people. They are alone allowed to read the Veda or Sacred Books. The Khatries or sect next in dignity, being only allowed to hear them read, while the other two read the Satras, or commentaries upon them; but the poor Chandalas are not allowed to enter their temple, or to be present at any religious ceremony.

In point of precedence, the Brahmins claim a superiority even to princes, the latter being chosen of Khatry or second sect. In fact the Brahmin claims every privilege, and the inferior sects give place to him; the Hindoos are allowed to eat no flesh nor to shed blood. Their food is rice and dholl, and other vegetables, dressed with ghee (dholl is a kind of split pea, ghee, a kind of butter, melted and refined to make it capable of being kept a long time) and seasoned with ginger and other spices. The food which they most esteem is milk, as coming from the cow; an animal for which they have the most extravagant veneration, insomuch that it is enacted in the code of Gentoo laws, that any one who exacts labour from a bullock that is hungry or thirsty, or shall oblige him to labour when fatigued, is liable to be fined by the magistrates.

The Hindoos are remarkable for their ingenuity in all kinds of handicraft; but their utensils are simple and in many respects inconvenient, so that incredible labour and patience are necessary, for the accomplishment of any work; and for this the Hindoos are remarkable. The religion of the Hindoos is contained in certain books, called Vedas; and, though now involved in superstition, seems to have been originally pure, inculcating the belief of an Eternal Being, possessed of every divine perfection. Their subordinate deities, Brahma, Vishnou, and Sheevah, are only representatives of the wisdom, goodness, and power of the supreme god Brahma; whom they call the principles of Truth, the spirit of Wisdom, and the Supreme Being; so that it is probable that all their idols were at first only designed to represent these attributes: they believe in ten Avators, or incarnation of the Deity, nine of which have taken place for the punishment of tyrants, or removing some great natural calamity; and the tenth is to take place at the dissolution of the universe. Several of the Avators inculcate the transmigration of souls, and the ninth of them, which forbids the sacrifices of animals, gave rise to the religion of Gauda Boodma, or Fo.

Their deities are extremely numerous, and are generally supposed to have first originated in Italy and Greece.

After stopping six days, I proceeded to Bombay, and on the 30th of May I arrived there. After delivering my passport, I made application for a ship for England, and was some time before I could get one; and the great expense I incurred in living at a tavern, made me entirely pennyless, so that I was forced to dispose of the shawls which I had presented me by the Rajah of Omrouty, and for which I received three hundred rupees each. But before I was finally settled, I had not above ten rupees left.

Bombay is an island of Hindostan, on the west coast of the Deccan, seven miles in length, and about twenty-one miles in circumference; the ground is barren, and good water scarce; it was formerly considered very unhealthy, but by draining the swamps and bogs the air is much improved; the inhabitants are of several nations and very numerous, but are principally Persians.

The religion of the Persians is, generally, Paganism, directed principally by the priests of magi, men of strict austere life, forbidding the use of either ornament or gold; making the ground their bed, and herbs their food. Their whole time is spent in offering to the gods the prayers and sacrifices of the people, as they only might be heard.

The people are *Gentiles*; as to their religion, they worship the sun and moon, and various heavenly bodies, from whom they suppose they derive every blessing of light and warmth; and every morning they gather themselves round the beech and present their morning oblations, by pouring into the sea quantities of milk and odoriferous flowers, and prostrating themselves with their faces to the earth, as a mark of adoration to their rising deity (the sun.) Besides other gods which the Gentiles worship, they are great idolaters of fire, which they offer sacrifices to in time of peace, and carry it with them, as their tutelar deity in time of war. Their adoration is so great, that the first candle they see lighted, let it be in whose place it will, they immediately stop and repeat a prayer. In their habitation they never put it out after it is once lighted.

Besides the town of Bombay, which is about a mile in length, with mean houses (a few only excepted), there is a capacious harbour or bay, reckoned the finest haven in the east, where all ships may find security from the inclemency of the different seasons. After remaining here for the space of three months, I was engaged as captain's clerk on board the Hon. Company's Ship Marquis of Huntly. We sailed from hence July 25, 1820, and arrived at the new anchorage in nineteen days' sail; soon after I went up to Calcutta on duty for the ship.

Calcutta, or *Fort William*, the emporium of Bengal, and principal seat of India, is situated on the western side of the Hoogely river, at about ninety-six miles from its mouth, which is navigable up to the town for large ships. This extensive and beautiful town is supposed to contain between four and five hundred thousand inhabitants. The houses are variously built, some of brick, others of mud and cow-dung, and a great number with bamboos (a large kind of reed or cane) and mats. The bamboos are placed as stakes in the ground, and crossed with others in different ways, so as to enable them to make the matting fast, when for the roofing they lay them one upon the other, when a large family lie in that small compass of about six feet square, which makes a very motley appearance. The mixture of European and Asiatic manners observed in Calcutta is wonderful; coaches, phaetons, hackeries, two-wheeled carriages drawn by bullocks, palanquins carried by the natives, and the passing ceremonies of Hindoos, and the different appearance of the faquirs, form a diversified and curious appearance. The European houses have, many of them, the appearance of palaces or temples, and the inhabitants are very hospitable.

After the cargo was sent on board I returned to the ship, but on our passage down the river we were compelled to lie out in the river, owing to the great boar, as it is called; it is a quick overflowing of the water, which rises in a great body and with such violence that it breaks down all before it. It arises from the narrowness of the river, and the force which it makes from the sea; in the course of two minutes it rises to the height of four or five feet.

Lying in one of the creeks till the tide was turned, I was greatly alarmed by the men getting into the boat in great disorder and telling me that it was a crocodile which I had for a long time observed, and mistaken for the hull of a tree. A crocodile is an amphibious voracious animal, in shape resembling a lizard. It is covered with very hard scales, which cannot but with difficulty be pierced, except under the belly, where the skin is tender. It has a wide throat, with several rows of teeth, sharp and separated, which enter one another.

On my arrival on board every thing was in confusion, as we expected to sail in a few days.

SECTION IX

THE AUTHOR'S DEPARTURE FROM CALCUTTA, AND ARRIVAL AT CHINA—AN ACCOUNT OF THEIR RELIGION, CUSTOMS, AND MANNERS, AND OF HIS BEING ROBBED ON DANES' ISLAND—THE AUTHOR'S DEPARTURE FROM CHINA AND ARRIVAL AT ANJURE POINT—THE CUSTOMS AND MANNERS OF THE MALAYS—DEPARTURE THEREFROM, AND ARRIVAL AT ST. HELENA—DESCRIPTION OF THE EMPEROR NAPOLEON'S TOMB AND HOUSES—DEPARTURE FROM ST. HELENA, AND ARRIVAL IN ENGLAND.

We sailed from Bengal in company with the Hon. Company's Ship Dunira, October 19th, 1820, with a fine breeze, and arrived at Pulo Penang, or Prince of Wales's Island, on the 6th of November. The houses have a noble appearance, and are built after the form of those in Calcutta. The inhabitants are principally Malays; of them I shall speak more hereafter. After having received on board a quantity of rattan, as private trade for the captain, we made sail and arrived at Macao, on January 26th, 1821, after a long and tedious voyage.

Macao, a town of China, in the province of Canton, is seated in an inland at the entrance of the river Tae. The Portuguese have been in possession of the town and harbour since the early part of the seventeenth century. The houses are low and built after the European manner; the Portuguese are properly a mixed breed, having been married to Asiatic women. Here is a Portuguese Governor as well as a Chinese Mandarin. The former nation pays a great tribute to choose their own magistrates. The city is defended by three forts, built upon eminences; and the works are good and well planted with artillery.

On the 29th we anchored off the second bar, and found lying here the Hon. Company's Ship Canning, and two or three other Company's ships; on the 30th weighed and made sail, but there not being water enough, removed back to our old station. On the following day we crossed Whampo. After the cargo was discharged I went up to Canton.

Canton is a large and populous city, situated in one of the first rivers in the empire. It is the capital of the province of Quan-tong, and the centre of the European trade in that country. The streets are long and straight, paved with flag stones, and adorned with lofty arches. The houses are remarkably neat, but consist only of one story, and they have no windows to the streets. The covered market place is full of shops. The inhabitants are estimated at about 1,000,000; many of whom reside in barks, which touch one another, form a kind of floating city, and are so arranged as to form streets. Each bark lodges a family and their grand children, who have no other dwelling. At break of day all the people who inhabit them depart to fish or to cultivate their rice.

The frugal and laborious manner in which the great live, the little attention which is paid to the vain and ridiculous prejudice of marrying below rank; the ancient policy of giving distinction to men and not to families, by attaching nobility only to employments and talents, without suffering it to be hereditary; and the decorum observed in public, are admirable traits in the Chinese character.

There is little distinction in the dress of men and women; rank and dignity are only distinguished by the ornaments they wear, and they dare not presume to wear any thing without proper authority, without being severely chastised for it. Their dress in general consists of a long vest, which reaches to the ground, one part of it, on the left side, folds over the other, and is fastened to the right by four or five small gold or silver buttons placed at a little distance from one another. The sleeves are wide towards the shoulder, and grow narrow towards the wrist—they terminate in the form of a horse-shoe—round their middle they wear a large girdle of silk, the ends of which hang down to their knees; from this girdle is suspended a sheath, containing a knife, and over all they wear a loose jacket down to the middle, with loose short sleeves, generally lined with fur, and under all they wear a kind of net to prevent it from chafing. The general colour of these dresses is black or blue.

Their religion is idolatry, their principal idol is *Fong Chon*, and they are very superstitious, believing in magic and invocation of spirits, and the art of foretelling events by divination.

While receiving our cargo on board, a Chinaman belonging to one of the craft, stole a box of tea, but, by the exertion of our officers, the culprit was taken and immediately sent on shore to Dane's Island to the mandarine. He was found guilty of the crime, and his punishment three dozen blows with the bastinado. The instrument of correction, called pan-tsee, is a bamboo a little flattened, broad at the bottom, and polished at the upper extremity, in order to manage it more easily with the hand.

The culprit, after the mandarin has given the signal for punishment, is seized and stretched out with his belly flat on the ground, his breeches are pulled down to his heels, and on the mandarine throwing down a stick, of which he has a number by him, one of the officers in attendance uses the pan-tsee, and gives him five severe blows, which are succeeded by several others till the number is complete. When it is over, the criminal must throw himself on his knees, incline his body three times to the earth, and thank his judge for the trouble he has taken in his correction.

The mandarins are of two classes, viz.; those of letters, and the inferior sort are styled mandarins of arms. The latter class do not enjoy the same consideration as the former.

The Chinese in general are much addicted to commit depredations on the pockets, or, in fact, on any unguarded property. After all our cargo was received on board, I went in company with two midshipmen, Mr. C—— and Mr. R——, on Dane's Island. After we landed some Chinese came and decoyed us to their village, which was at the back of a number of hills and out of sight of the shipping, under a promise that they would let us have some of their country fruit, such as they sent us on board. The length of time that some of them were absent, and the sun going down fast, made us rather doubt the sincerity of their intentions; those that were with us begged that we would stop till the sun was down, but we began to be afraid of our lives. When the men saw that we were determined to wait no longer, they gave a dreadful whoop, which was answered by others stationed on the hills; they immediately seized hold of us and rifled our pockets.

On March 25th we sailed down to Macao, and on the following day we took our departure, and on the 24th of April arrived at Anjier point, and is a settlement belonging to the Dutch; it lies to the east of Batavia. The houses are generally built of bamboo; the inhabitants are of various casts, Pagans, Mahometans, and Chinese. The barbarism of the Batta Tribes is horrible, for they kill and eat their criminals or prisoners of war, or even sacrifice their own relations when aged and infirm, not so much with a view to gratify their appetites, as to perform a pious ceremony. Thus, when a man becomes infirm and weary of the world, he is said to invite his own children to eat him when salt and limes are cheapest. He then ascends a tree, round which his friends and offspring assemble, and as they shake the tree they join in a funeral dirge, the import of which is, the season is come, the fruit is ripe, and it must descend. The victim descends, and those that are nearest deprive him of life, and devour his remains in a solemn banquet.

In a few days we made sail. We arrived at St. Helena, on the 10th of July, 1821. This island is situated in the South Atlantic Ocean, its circumference is about twenty miles, and at a distance it has the appearance of a large rock rising out of the sea. On rounding the island it has a very romantic appearance; the town lying in a valley presents to the eye a beautiful chain of scenery. It has some very high mountains, particularly one called Diana's Peak, which is covered with wood to the very summit. There are other hills also, which bear a volcanic appearance, and some have huge rocks of lava, and a kind of half-vitrified flags. James Town is erected in a valley at the bottom of a bay, between two steep dreary mountains, and has from the shipping a noble appearance.

Accommodations are tolerably good, and the inhabitants, generally speaking, are very hospitable. Their villas are pleasantly situated, and have a fine view of the sea; the whole face of the country is really romantic; the hills are immensely high, and the valleys very narrow; and in many of them there are a few houses, which give the whole island a very picturesque appearance.

After obtaining a passport from the Adjutant-General, I went over a long succession of hills to see the habitations of the late Emperor Napoleon Bonaparte. The roads were very difficult to ascend, and particularly rugged. The remains of this great and illustrious personage are buried in a deep valley, about three miles from James town, and about two miles from his late residence at Longwood, under the peaceful shade of three weeping willows, and which also, (as in respect to his dust,) lend a solemn air of reverential darkness to the memorable *well*, from which, during his pilgrimages, he was wont to receive his refreshing draughts.

No stately monument marks the spot; no polished alabaster, or the mimicry of sculptured marble marks his grave: the real excellency of the patriot is written on the minds of his countrymen; it will be remembered with applause as long as the nation subsists, without this artificial expedient to perpetuate it.

Let the poor pass by his grave, and thankfully acknowledge, there lies the man who gloriously fought for his country and his subjects, to free them from the galling yoke of tyranny and oppression: no tablets are written to mark his actions, but those which are written in the heart of his subjects.

The depth of his tomb is about twenty feet, and his coffin rests upon two pedestals, ten feet high. His body is enclosed in four coffins, first lead, second deal, third mahogany, and fourth marble. What is very remarkable is, that part of his tomb is made of the flag-stones of his new house, taken out of one of the kitchens. After viewing the tomb of the man who was the most brilliant meteor in the political world, I proceeded up to Longwood, to take a view of the habitation in which he died.

After presenting my passport I had permission to inspect the premises: the officer took great pains in shewing me the very spot on which he quitted his troubles and persecutions, when he kindly left me to make what sort of reflections I thought proper. The darkness of the room gave it a very solemn appearance, and suited the mind to contemplate upon this late extraordinary character;—but a short period past he was the terror of the world, and now, alas! what is he? He is laid low in the tomb, unregretted and unpitied by his merciless enemies. A gleam of light through the casements reflected a dead glimmer through the gloomy mansion. The *most illustrious* have claimed the *tomb* for their last retreat; rooms of state are resigned! the sceptre has ceased to wield, and sumptuous banquets are neglected for no other ornament than the winding sheet! "Where is the star that blazed upon his breast, or the coronet that glittered round his temples?" Alas! they are resigned and given over, through the power of the tyrant hand of death.

I have often walked between the impending promontory's craggy cliff; I have sometimes trod the vast spaces of the lonely desert, and penetrated the inmost recesses of the dreary cavern; but never beheld Nature lowering with so dreadful a form; never felt impressions of such awe striking cold on my heart, as under this roof; every thing seemed to participate in grief for their deceased lord. The rooms were very dirty and much neglected. The plants in his late garden seemed to droop their heads in sorrow for the loss of the hand that reared them.

I next proceeded to the palace which had been sent from England, and really it would have reflected honour on the British nation, and no sovereign in the world need wish for a more magnificent one, had it been placed in a more healthy part of the island.

We sailed for England on the 29th, and arrived on the 13th of September, 1821, after a speedy and pleasant passage.